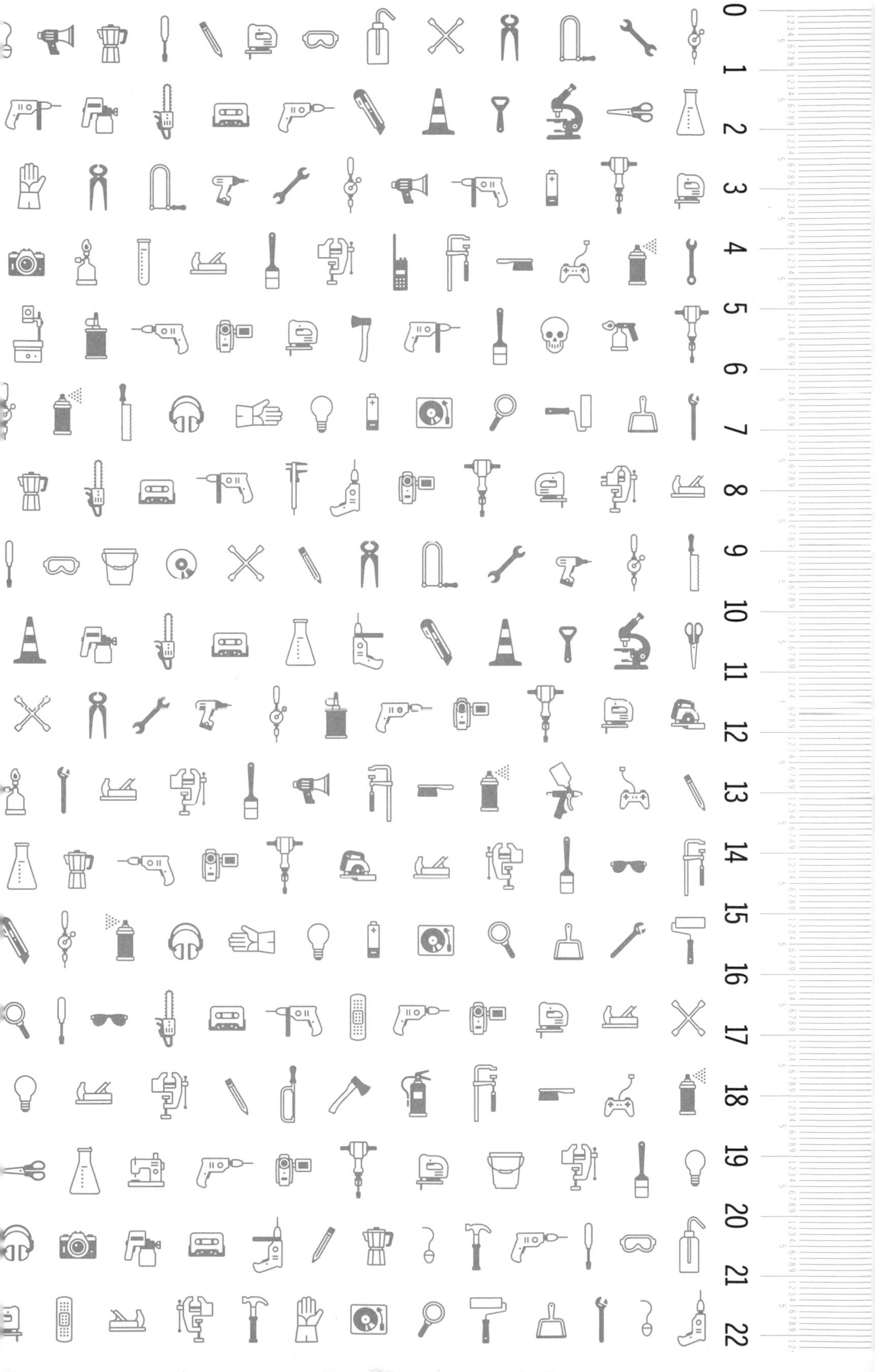
0
1
2
3
4
5
6
7
8
9
10
11
12
13
14
15
16
17
18
19
20
21
22

Maker's Notebook
by the Staff of MAKE Magazine

Published by Make:Books, an imprint of Maker Media, a division of O'Reilly Media, Inc.,
1005 Gravenstein Highway North, Sebastopol, CA 95472.

O'Reilly books may be purchased for educational, business, or sales promotional use. For
more information, contact our corporate/institutional sales department: 800–998–9938
or corporate@oreilly.com.

Print History: April 2008: First Edition

Publisher: Dale Dougherty
Associate Publisher and Executive Editor: Dan Woods
Editor: Gareth Branwyn
Creative Director: Daniel Carter
Art Director: Brian Scott/Boon
Production Manager: Terry Bronson
Illustrator: Damien Scogin
Content: Gareth Branwyn, Terry Bronson, Paul Spinrad

ISBN-10: 0-596-51941-9
ISBN-13: 978-0-596-51941-4

IDEA/PROJECT

DATE

NOTES/SIG

FROM PAGE

TO PAGE

IDEA/PROJECT

DATE

NOTES/SIG

FROM PAGE

TO PAGE

IDEA/PROJECT

DATE

NOTES/SIG

FROM PAGE

TO PAGE

IDEA/PROJECT

DATE

NOTES/SIG

FROM PAGE

TO PAGE

IDEA/PROJECT

DATE

NOTES/SIG

FROM PAGE

TO PAGE

IDEA/PROJECT

DATE

NOTES/SIG

FROM PAGE

TO PAGE

IDEA/PROJECT

DATE

NOTES/SIG

FROM PAGE

TO PAGE

IDEA/PROJECT

NOTES/SIG

DATE

FROM PAGE

TO PAGE

IDEA/PROJECT

DATE

NOTES/SIG

FROM PAGE

TO PAGE

IDEA/PROJECT

DATE

NOTES/SIG

FROM PAGE

TO PAGE

IDEA/PROJECT

DATE

NOTES/SIG

FROM PAGE

TO PAGE

IDEA/PROJECT

DATE

NOTES/SIG

FROM PAGE

TO PAGE

PROJECT

DATE

S/SIG

FROM PAGE

TO PAGE

PROJECT

/SIG

DATE

FROM PAGE

TO PAGE

103

PROJECT

SIG

DATE

FROM PAGE

TO PAGE

105

PROJECT

SIG

DATE

FROM PAGE

TO PAGE

107

PROJECT

/SIG

DATE

FROM PAGE

TO PAGE

109

PROJECT

SIG

DATE

FROM PAGE

TO PAGE

111

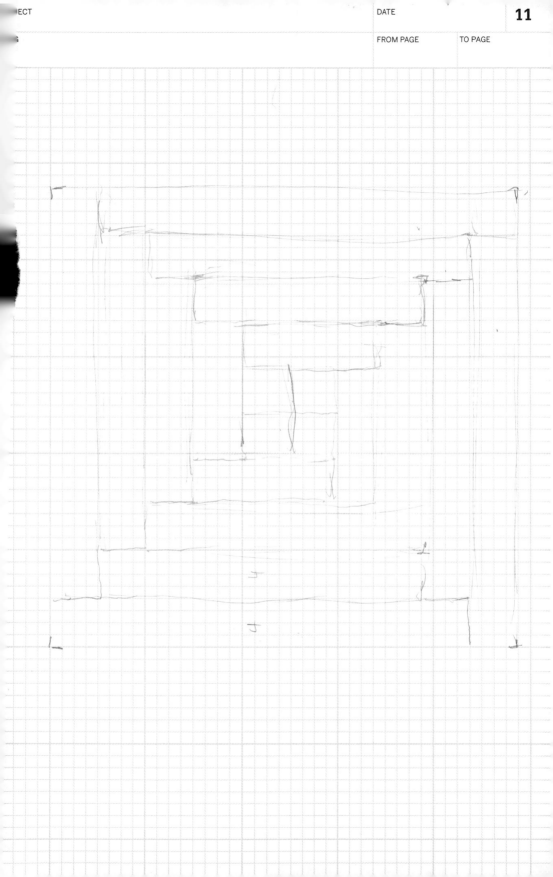

IDEA/PROJECT

DATE

NOTES/SIG

FROM PAGE

TO PA

PROJECT

SIG

DATE

FROM PAGE

TO PAGE

113

PROJECT

SIG

DATE

FROM PAGE

TO PAGE

115

PROJECT

SIG

DATE

FROM PAGE

TO PAGE

117

PROJECT

/SIG

DATE

FROM PAGE

TO PAGE

123

PROJECT

/SIG

DATE

FROM PAGE

TO PAGE

127

PROJECT

S/SIG

DATE

FROM PAGE

TO PAGE

129

PROJECT

/SIG

DATE

FROM PAGE

TO PAGE

131

PROJECT

S/SIG

DATE

FROM PAGE

TO PAGE

133

PROJECT

DATE

135

S/SIG

FROM PAGE

TO PAGE

PROJECT

SIG

DATE

FROM PAGE

TO PAGE

139

PROJECT

SIG

DATE

FROM PAGE

TO PAGE

141

PROJECT

SIG

DATE

FROM PAGE

TO PAGE

143

Reference

THE MAKER'S BILL OF RIGHTS

■ Meaningful and specific parts lists shall be included. ■ Cases shall be easy to open. ■ Batteries should be replaceable. ■ Special tools are allowed only for darn good reasons. ■ Profiting by selling expensive special tools is wrong and not making special tools available is even worse. ■ Torx is OK; tamperproof is rarely OK. ■ Components, not entire sub-assemblies, shall be replaceable. ■ Consumables, like fuses and filters, shall be easy to access. ■ Circuit boards shall be commented. ■ Power from USB is good; power from proprietary power adapters is bad. ■ Standard connecters shall have pinouts defined. ■ If it snaps shut, it shall snap open. ■ Screws better than glues. ■ Docs and drivers shall have permalinks and shall reside for all perpetuity at archive.org. ■ Ease of repair shall be a design ideal, not an afterthought. ■ Metric or standard, not both. ■ Schematics shall be included.

Drafted by Mister Jalopy, with assistance from Phillip Torrone and Simon Hill.

CRAFTER'S MANIFESTO

■ People get satisfaction for being able to create/craft things because they can see themselves in the objects they make. This is not possible in purchased products. ■ The things that people have made themselves have magic powers. They have hidden meanings that other people can't see. ■ The things people make they usually want to keep and update. Crafting is not against consumption. It is against throwing things away. ■ People seek recognition for the things they have made. Primarily it comes from their friends and family. This manifests as an economy of gifts. ■ People who believe they are producing genuinely cool things seek broader exposure for their products. This creates opportunities for alternative publishing channels. ■ Work inspires work. Seeing what other people have made generates new ideas and designs. ■ Essential for crafting are tools, which are accessible, portable, and easy to learn. ■ Materials become important. Knowledge of what they are made of and where to get them becomes essential. ■ Recipes become important. The ability to create and distribute interesting recipes becomes valuable. ■ Learning techniques brings people together. This creates online and offline communities of practice. ■ Craft-oriented people seek opportunities to discover interesting things and meet their makers. This creates marketplaces. ■ At the bottom, crafting is a form of play.

Drafted by Ulla-Maaria Mutanen, creator of the HobbyPrincess blog.

Common Weights and Measures

Length

1 inch	1/36 yard	1/12 foot
1 yard	3 feet	
1 rod	5 ½ yards	
1 furlong	220 yards	40 rods
1 mile	1,760 yards	5,280 feet
1 fathom	6 feet	
1 nautical mile	6,076.1 feet	

Length (Metric System)

1 millimeter	1/1000 meter
1 centimeter	1/100 meter
1 decimeter	1/10 meter
1 meter (basic unit of length)	
1 dekameter	10 meters
1 kilometer	1,000 meters

Area

1 square inch	1/1296 square yard	1/144 square foot
1 square yard (basic unit of area)		
1 square rod	30 ¼ square yards	
1 acre	4,840 square yards	160 square rods

Volume (Dry & Liquid)

1 cubic inch	1/46656 cubic yard	1/1728 cubic foot
1 cubic yard (basic unit of volume)		
1 U.S. fluid oz.	1/128 U.S. gallon	1/16 U.S. pint
1 pint	⅛ gallon	½ quart
1 U.S. gallon	231 cubic inches	
1 dry pin	1/64 bushel	½ dry quart
1 dry quart	1/32 bushel	⅛ peck
1 peck	¼ bushel	
1 U.S. bushel	2,150.4 cubic inches	

Weight (Avoirdupois)

1 grain	1/7000 pound	1/437.5 ounce
1 dram	1/256 pound	1/16 ounce
1 ounce	1/16 pound	
1 pound	16 ounces	
1 short hundred wt.	100 pounds	
1 long hundred wt.	112 pounds	
1 short ton	2,000 pounds	
1 long ton	2,240 pounds	

Conversion Calculations

To change	To	Multiply by:
centimeters	inches	0.3937
centimeters	feet	0.03281
cubic feet	cubic meters	0.0283
cubic meters	cubic feet	35.3145
cubic meters	cubic yards	1.3079
cubic yards	cubic meters	0.7646
degrees	radians	0.01745
feet	meters	0.3048
feet	miles (statute)	0.0001894
feet/second	miles/hour	0.6818
gallons (U.S.)	liters	3.7853
grams	ounces (avdp)	0.0353
grams	pounds	0.002205
horsepower	watts	745.7
horsepower	Btu/hour	2,547
hours	days	0.04167
inches	millimeters	25.4
inches	centimeters	2.54
kilograms	pounds	2.2046
kilometers	miles	0.6214
kilowatt-hour	Btu	3412
liters	gallons (U.S.)	0.2642
liters	pints (dry)	1.8162
liters	pints (liquid)	2.1134
liters	quarts (dry)	0.9081
liters	quarts (liquid)	1.0567
meters	feet	3.2808
meters	miles	0.0006214
meters	yards	1.0936
miles	kilometers	1.6093
miles	feet	5280
millimeters	inches	0.0394
ounces	pounds	0.0625
pounds (avdp)	kilograms	0.4536
pounds	ounces	16
quarts (dry)	liters	1.1012
quarts (liquid)	liters	0.9463
square feet	square meters	0.0929
square kilometers	square miles	0.3861
square meters	square feet	10.7639
square miles	square kilometers	2.59
square yards	square meters	0.8361
watts	Btu/hour	3.4121
watts	horsepower	0.001341
yards	meters	0.9144
yards	miles	0.0005682

Classic Design Tradeoffs

Time/Space, Compression/Quality, Cost/Time, Speed/Accuracy, Convenience/Privacy, Reliability/Maintenance Time, Simplicity/Flexibility, Security/Flexibility, Security/Freedom, Power/Efficiency, Scalability/Performance, Resolution/Overhead, Bandwidth/Latency.

Conversions

Fractions	Decimals
½	0.5000
⅓	≈ 0.3333
¼	0.2500
⅕	0.2000
⅙	≈ 0.1667
⅐	≈ 0.1429
⅛	0.1250
⅑	≈ 0.1111
¹⁄₁₀	0.1000
¹⁄₁₁	≈ 0.0909
¹⁄₁₂	≈ 0.0833
¹⁄₁₆	0.0625
¹⁄₃₂	≈ 0.0313
¹⁄₆₄	≈ 0.0156
⅔	≈ 0.6667
⅖	0.4000
²⁄₇	≈ 0.2857
²⁄₉	≈ 0.2222
²⁄₁₁	≈ 0.1818
¾	0.7500
⅗	0.6000
³⁄₇	≈ 0.4286
⅜	0.3750
³⁄₁₀	0.3000
³⁄₁₁	≈ 0.2727
⅘	0.8000
⁴⁄₇	≈ 0.5714
⁴⁄₉	≈ 0.4444
⁴⁄₁₁	≈ 0.3636
⅚	≈ 0.8333
⁵⁄₇	≈ 0.7143
⅝	0.6250
⁵⁄₉	≈ 0.5556
⁵⁄₁₁	≈ 0.4545
⁵⁄₁₂	≈ 0.4167
⁶⁄₇	≈ 0.8571
⁶⁄₁₁	≈ 0.5455
⅞	0.8750
⁷⁄₉	≈ 0.7778
⁷⁄₁₀	0.7000
⁷⁄₁₁	≈ 0.6364
⁷⁄₁₂	≈ 0.5833
⁸⁄₉	≈ 0.8889
⁸⁄₁₁	≈ 0.7273
⁹⁄₁₀	0.9000
⁹⁄₁₁	≈ 0.8182
¹⁰⁄₁₁	≈ 0.9091
¹¹⁄₁₂	≈ 0.9167

Length: Metric Conversions	
1 centimeter	0.39 inch
1 inch	2.54 centimeters
1 meter	39.37 inches
1 foot	0.305 meter
1 meter	3.28 feet
1 yard	0.914 meter
1 meter	1.094 yards
1 kilometer	0.62 mile
1 mile	1.609 kilometers

Volume: Metric Conversions	
1 cubic centimeter	0.06 cubic inch
1 cubic inch	16.4 cubic centimeters
1 cubic yard	0.765 cubic meter
1 cubic meter	1.3 cubic yards
1 milliliter	0.034 fluid ounce
1 fluid ounce	29.6 milliliters
1 U.S. quart	0.946 liter
1 liter	1.06 U.S. quarts
1 U.S. gallon	3.8 liters
1 liter	0.9 dry quart
1 dry quart	1.1 liters

	°Fahrenheit	°Celsius
Boiling point of water	212°	100°
Freezing point of water	32°	0°
Absolute zero	−459.6°	−273.1°

Caffeine

Substance	Ounces	Milligrams
Cocaine Energy Drink	8.4 oz	280 (33.3/oz)
Jolt	23.5 oz	280 (11.9/oz)
SoBe No Fear Super Energy	16 oz	174/(10.9/oz)
Monster Energy	16 oz	160/(10/oz)
Rockstar	16 oz	160/(10/oz)
Full Throttle	16 oz	144/(9/oz)
Yerba Mate, traditional	6oz	110
Coffee	8 oz	65–120
Espresso shot	1 oz	65–130
Foosh Energy Mint		1: 100
Red Bull	8.3 oz	80/(9.16/oz)
Mountain Dew	12 oz	54
Tea, green or black	8 oz	50
Yerba Mate tea	8 oz	50
Diet Coke, Tab	12 oz	47
Dr Pepper	12 oz	41
Sunkist	12 oz	41
Pepsi	12 oz	38
Coca-Cola Classic	12 oz	35

U.S. City Area Codes

City	State	Area Code
Birmingham	AL	205
Mobile	AL	251
Montgomery	AL	334
Little Rock	AR	501
Scottsdale	AZ	480
Tucson	AZ	520
Phoenix	AZ	602
Flagstaff	AZ	928
Los Angeles	CA	213
San Francisco	CA	415
San Jose	CA	408
Oakland	CA	510
San Diego	CA	619
Sebastopol (Home of MAKE & CRAFT)	CA	707
Sacramento	CA	916
Denver	CO	303/720
Colorado Springs	CO	719
New Haven	CT	203
Washington	DC	202
Delaware: entire state	DE	302
Miami	FL	305/786
Orlando	FL	407/321
Tampa	FL	813
Atlanta	GA	404
Guam	GU	671
Hawaii: entire state	HI	808
Cedar Rapids	IA	319
Des Moines	IA	515
Idaho: entire state	ID	208
Chicago	IL	312/773
Springfield	IL	217
Fort Wayne	IN	260
Indianapolis	IN	317
Wichita	KS	316
Kansas City	KS	913
Lexington	KY	859
New Orleans	LA	504
Shreveport	LA	318
Boston	MA	617
Baltimore	MD	410/443
Maine: entire state	ME	207
Detroit	MI	313
Grand Rapids	MI	616
Minneapolis	MN	612
St. Paul	MN	651
St. Louis	MO	314
Kansas City	MO	816
Biloxi	MS	228
Montana: entire state	MT	406
Charlotte	NC	704/980
Greensboro	NC	336
North Dakota: entire state	ND	701
Omaha	NE	402
New Hampshire: entire state	NH	603
Trenton	NJ	609
Newark	NJ	973/862
New Mexico: entire state	NM	505
Las Vegas	NV	702
Reno	NV	775
New York City	NY	212/646/917
Long Island	NY	516/631
Cleveland	OH	216
Cincinnati	OH	513
Oklahoma City	OK	405
Portland	OR	503/971
Philadelphia	PA	215/267
Pittsburgh	PA	412/878
Puerto Rico: entire territory	PR	787/939
Rhode Island: entire state	RI	401
Charleston	SC	843
South Dakota	SD	605
Nashville	TN	615
Memphis	TN	901
Dallas	TX	214
Austin	TX	512
Salt Lake City	UT	801
U.S. Virgin Islands: entire territory	VI	340
Arlington	VA	703/571
Richmond	VA	804
Vermont: entire state	VT	802
Seattle	WA	206
Spokane	WA	509
Milwaukee	WI	414
Madison	WI	608
West Virginia: entire state	WV	304
Wyoming: entire state	WY	307

Canada

City	Province	Area Code
Vancouver	B.C.	604/778
Calgary	Alberta	403
Toronto	Ontario	416/647
Ottawa	Ontario	613
Montréal	Québec	514

International Codes

Country	Code
Afghanistan	93
Argentina	54
Australia	61
Austria	43
Belgium	32
Brazil	55
Bulgaria	359
Canada	1
Central African Republic	236
Chile	56
China, People's Republic	86
Czech Republic	420
Denmark	45
Egypt	20
Finland	358
France	33
Germany	49
Greece	30
Hong Kong	852
Hungary	36
India	91
Iran	98
Iraq	964
Ireland	353
Israel	972
Italy	39
Jamaica	1876
Japan	81
Korea (South)	82
Kuwait	965
Luxembourg	352
Macau	853
Martinique	596
Mexico	52
Netherlands	31
New Zealand	64
Nigeria	234
Norway	47
Pakistan	92
Palestinian Settlements	970
Panama	507
Peru	51
Philippines	63
Poland	48
Portugal	351
Romania	40
Russia	7
Saudi Arabia	966
Singapore	65
Solomon Island	677
South Africa	27
Spain	34
Sweden	46
Switzerland	41
Syria	963
Taiwan	886
Thailand	66
Turkey	90
United Arab Emirates	971
United Kingdom	44
United States of America	1
U.S. Virgin Islands	1–340
Vatican City	39/379
Venezuela	58

Common English <-> 1337 Character Substitutions

A 4, B 8, E 3, I |, L 1, O 0, S $, T 7, Z 2

Mnemonic Devices

(Memory aids for Makers)

MEASUREMENT
Device: Knuckles and dips
For: Remembering months. With your fists together, start with the first knuckle and go across, saying the months for each knuckle and dip. A knuckle is 31 days, a dip 30 (except February 28/29).

Device: King Hector died mysteriously drinking chocolate milk.
For: Metric system measurements (greatest to least): Kilo, Hecto, Deka, Meter, Dec, CM, MM

MATHEMATICS
Device: How I need a drink, alcoholic of course, after the heavy chapters involving geodesy.
For: The numbers of Pi (by counting the letters) to 14 places: 3.1415926535897

Device: The pinch goes to the smaller number.
For: remembering less than (<) and greater than (>)

SCIENCE
Device: Kids prefer cheese over fried green spinach.
For: Order of taxonomies in biology: Kingdom, Phylum, Class, Order, Family, Genus, Species

Device: Environment is ABC
For: Remembering the parts of an environment: Abiotic (non-living), Biotic (living), and Cultural (human-made)

Device: Anyone can make pretty high heeled shoes
For: Classification of humans: Animalia, Chordata, Mamalia, Primata, Hominadae, Homo, Sapien

Device: Mercury viewed Earth's many aspects joyfully sitting under Neptune
For: Remembering the order of planets (and Asteroid Belt) from the Sun: Mercury, Venus, Earth, Mars, Asteroid Belt, Jupiter, Saturn, Uranus, Neptune

Device: OIL RIG
For: Oxidation Is Loss (of electrons), Reduction Is Gain (of electrons)

TECHNOLOGY
Device: Roy G. Biv
For: The visible electromagnetic spectrum: Red, Orange, Yellow, Green, Blue, Indigo, Violet

Device: Raging Martians invaded Roy G. Biv using x-ray guns
For: Remembering waves of the electromagnetic spectrum from longest to shortest: Radio, microwave, infrared, visible, ultraviolet, x-ray, gamma ray

Device: Please do not throw sausage pizza away
For: Order of layers in the Open System Interconnection (OSI) computer network protocol:
Physical Layer, Data Link Layer, Network Layer, Transport Layer, Session Layer, Presentation Layer, Application Layer

Device: Black beetles running on your garden bring very good weather
For: Order of resistor color bands: Black, Brown, Red, Orange, Yellow, Green, Blue, Violet, Gold, White

Device: Twinkle Twinkle Little Star Power equals I squared R
For: Power (watts) = Volts2/Resistance (Ohms)

Device: Good models know how Dunkin Donuts can make µ not petite
For: The ordering of common Greek size prefixes (from largest to smallest): giga, mega, kilo, hecto, deca, deci, centi, milli, micro (µ), nano, pico

Device: I feel rather negatively about cats
For: Remembering that cathode is negative. Used by Dave Hrynkiw's (of Solarbotics.com) high school Physics teacher.

International Morse Code

A	• −	U	• • −
B	− • • •	V	• • • −
C	− • − •	W	• − −
D	− • •	X	− • • −
E	•	Y	− • − −
F	• • − •	Z	− − • •
G	− − •		
H	• • • •	0	− − − − −
I	• •	1	• − − − −
J	• − − −	2	• • − − −
K	− • −	3	• • • − −
L	• − • •	4	• • • • −
M	− −	5	• • • • •
N	− •	6	− • • • •
O	− − −	7	− − • • •
P	• − − •	8	− − − • •
Q	− − • −	9	− − − − •
R	• − •	[period/fullstop]	• − • − • −
S	• • •	[comma]	− − • • − −
T	−	[query]	• • − − • •

Hello, World!
In Various Languages

APL	'Hello, world!'
BASIC	100 PRINT "Hello, World!" 110 END

C++	#include <iostream>
	int main()
	{
	  std::cout << "Hello, world!\n";
	}

Java	public class HelloWorld
	{
	  public static void main(String args[])
	  {
	    System.out.println("Hello, World");
	  }
	}

Lisp	(print "Hello, World!")
Perl	print "Hello, World!\n";
PHP	<?php echo "Hello, World!\n"; ?>
Python	print "Hello, World!"
Ruby	ruby -e ' puts "Hello, World!" '
sh	echo "Hello, world!"
TCL	puts "Hello, World!"

Sewing Machine
Needle Sizes

	American	European
Lighter	8	60
	9	65
	10	70
	11	75
	12	80
	14	90
	16	100
	18	110
Heavier	19	120

Choose a size 8/60 needle for lightweight fabric, a 10/70 for medium-weight material, a 14/90 or 16/100 for heavy fabrics like jeans, upholstery, canvas, etc., and 18/110 or 19/120 for the heaviest fabrics.

Common Technical Abbreviations

A	ampere (also "amp")	MOSFET	metal-oxide-semiconductor field-effect transistor
AAC	Apple audio codec	MUX	multiplex (or "multi-user experience")
ABS	acrylonitrile butadiene styrene	N	newton (unit of force)
A/D	analog-to-digital	NC	normally closed (also "no contact")
ADC	analog-to-digital conversion (or "ADC")	NiMH	nickel metal hydride
Ah	ampere-hour	NDA	non-disclosure agreement
ASCII	American Standard Code for Information Interchange	NIH	not invented here
ASIC	application-specific integrated circuit	NIMBY	not in my backyard
AWG	American Wire Gauge	NIST	National Institute of Standards and Technology
BFO	beat frequency oscillator	NO	normally open
BGA	ball grid array	NPN	negative-positive-negative
BJT	bipolar junction transistor	N-s	newton-second (unit of impulse)
BOM	bill of materials	OpAmp	operational amplifier
BP	bandpass	OSC	oscillator
BTU	British Thermal Unit	PCB	printed circuit board
C	coulomb (also "common" and "collector")	PCM	pulse-code modulation
CAP	capacitor	pF	picofarad
CCD	charge-coupled device	PNG	portable network graphics
CDMA	code-division multiple access	PNP	positive-negative-positive
CMOS	complementary metal-oxide semiconductor	POT	potentiometer
CNC	computer numerical control	PV	photovoltaic
COB	chip on board (or "close of business")	PWM	pulse-width modulation
CW	continuous wave	PZ	piezoelectric
DARPA	Defense Advanced Research Projects Agency	QR	Quick Response code
D/A	digital-to-analog	Qtz	quartz
DAC	digital-to-analog conversion (or "DAC")	R	resistance
DHCP	Dynamic Host Configuration Protocol	RC	resistor-capacitor
DIP	double in-line package	RFI	radio frequency interference
DoF	degrees of freedom (or "depth of field")	RSS	Real Simple Syndication (also "Rich Site Summary")
DPDT	double pole double throw	RTS	request to send
DRM	digital rights management	RXD	received exchange data
DSP	digital signal processing	SAE	Society of Automobile Engineers International
EEPROM	electrically erasable programmable read-only memory	SDK	software development kit
EIA	Electronics Industries Alliance	SEO	search engine optimization
EMF	electromotive force (or "electromagnetic field")	SI	International System of Units
EMI	electromagnetic interference	SIP	single in-line package
EOM	end of message	SMD	surface-mounted devices
ESD	electrostatic discharge	SMT	surface-mounted technology
F	farad (also "frequency")	S/N	signal-to-noise ratio
FET	field-effect transistor	SPST	single pole single throw
FLED	flashing LED	SQL	Software Query Language
FPGA	field-programmable gate arrays	S/S	stainless steel
FREQ	frequency	SSH	Secure Shell
FrieNDA	friendly non-disclosure agreement	SWG	Standard Wire Gauge
Gnd	Ground (voltage level)	T	tesla (unit of flux density)
H	henry	TLA	three-letter acronym
HF	high frequency	TO	transistor outline package
HV	high voltage	TTL	transistor-transistor logic
Hz	hertz	Tx	transmit
I	current (actually stands for "intensity")	TXD	transmit exchange data (see "RXD")
IEEE	Institute of Electrical and Electronic Engineers	UART	universal asynchronous receiver transmitter
IrDA	Infrared Data Association	UHF	ultra high frequency
IRQ	interrupt request	UPS	uninterrupted power supply
ISO	International Organization for Standardization	USS	United States Standard
J	joule	V	volt
JFET	junction field-effect transistor	Vcc	main supply voltage (power)
JTAG	Joint Test Action Group	Vdd	secondary supply voltage (power)
KWH	kilowatt hour	Vee	negative supply voltage (power)
LDR	light-dependent resistor	Vss	negative supply voltage (power)
LED	light-emitting diode	VFO	variable-frequency oscillator
LP	low-pass	VOM	volt-ohm meter (or multimeter)
mA	milliamperes	XMTR	transmitter
mcd	microcandela	YMMV	your mileage may vary
MCU	microcontroller unit (also "uC" or "μC")	Z	impedance
MOS	metal-oxide-semiconductor		

Web References

GETTING ORGANIZED/LIFE HACKING
43 Folders (43folders.com)
Creative Commons (creativecommons.org)
Getting Started with GTD (gtd.43folders.com)
Getting Things Done (davidco.com)
LifeClever (lifeclever.com)
LifeDev (lifedev.net)
Lifehacker (lifehacker.com)
Mentat Wiki (ludism.org/mentat)

JUICE BARS (CREATIVITY LINKS)
Creativity Portal (creativity-portal.com)
Creative Think (blog.creativethink.com)
Oblique Strategies (rtqe.net/ObliqueStrategies)
Notcot (notcot.org)
Mind Mapping (peterrussell.com/MindMaps)
TED (ted.com)

BUILDING PROJECTS/DIY
Afrigadget (afrigadget.com)
Craft: Blog (craftzine.com/blog)
Evil Mad Scientist Laboratories (evilmadscientistlabs.com)
Finkbuilt (finkbuilt.com)
Hackszine (hackszine.com)
Instructables (instructables.com)
Hooptyrides (hooptyrides.com)
Make: Blog (makezine.com/blog)
Music from Outer Space (musicfromouterspace.com)
ReadyMade (readymademag.com)
Steampunk Labs (steampunklabs.com)
uC Hobby (uchobby.com)

TECHNICAL REFERENCE
Bolt Depot (boltdepot.com/fastener-information)
Chip Directory (fer.nu/chipdir)
Circuit Symbols Library (tinyurl.com/25nwnz)
Datasheet Archive (datasheetarchive.com)
Electronics Conversions Formulas & References
(rfcafe.com/references/electrical.htm)
Lindsay Publications (lindsaybks.com)
Sci.Electronics.Repair FAQ (repairfaq.org)
The Hardware Book (hardwarebook.info)
This to That (Glue database) (thistothat.com)
Toolmonger (toolmonger.com)

FUN STUFF/DISTRACTIONS
Archie McPhee (mcphee.com)
Boing Boing (boingboing.net)
Damn Interesting (damninteresting.com)
Dark Roasted Blend (darkroastedblend.com)
Proceedings of the Athanasius Kircher Society (kirchersociety.org)
ScienceHack (sciencehack.com)
XKCD (xkcd.com)

Flickr Groups

MAKE (flickr.com/groups/make)
CRAFT (flickr.com/groups/craft)
Moleskinerie (flickr.com/groups/moleskinerie)
Moleskine Organization (flickr.com/groups/51462541@N00)
Notebook (flickr.com/groups/notebooks)

When I'm working on a problem, I never think about beauty, I only think about how to solve the problem. But when I have finished, if the solution is not beautiful, I know it is wrong.
— *R. Buckminster Fuller*

Suppliers/Surplus

SUPPLIERS
Adafruit Industries (adafruit.com)
BG Micro (bgmicro.com)
Digi-Key (digikey.com)
Jameco (jameco.com)
Mouser (mouser.com)
Newark (newark.com)
PartStore (partstore.com)
Ramsey (ramseyelectronics.com)
Small Parts (smallparts.com)
Solarbotics (solarbotics.com)
Sparkfun (sparkfun.com)
Tap Plastics (tapplastics.com)

SURPLUS SOURCES
All Electronics (allelectronics.com)
American Science and Surplus (sciplus.com)
Electronic Goldmine (goldmine-elec-products.com)
Herbach and Rademan (herbach.com)
Marlin P. Jones (mpja.com)
Surplus Tarders (www.73.com)
Weird Stuff Warehouse (weirdstuff.com)

Robot Laws, Rules, and Recipes

Asimov's Three (plus) Laws of Robotics
0. A robot may not injure humanity, or through inaction, allow humanity to come to harm.
1. A robot may not harm a human being, or, through inaction, allow a human being to come to harm.
2. A robot must obey the orders given to it by the human beings, except where such orders would conflict with the Zeroth or First Law.
3. A robot must protect its own existence, as long as such protection does not conflict with the Zeroth, First, or Second Law.

Tilden's Laws of Robotics
1. A robot must protect its existence at all costs.
2. A robot must obtain and maintain access to a power source.
3. A robot must continually search for better power sources.

Also known as:
1. Protect thy ass.
2. Feed thy ass.
3. Move thy ass to better real estate.

The Rodney Brooks Research Heuristic
Look for what is so obvious to everyone else that it's no longer on their radar, and put it on yours. Seek to uncover assumptions so implicit they're no longer being questioned. Question them. [Used by MIT AI Lab Director and iRobot founder Rodney Brooks.]

The Kenny Rogers Rule
When a project build turns frustrating, ugly; when the cursing starts. Step away. Take a break. It almost never fails. Corollary: The extent to which you don't want to stop is inversely proportional to the extent to which you need to. Why "Kenny Rogers?" Cause "you got to know when to hold, know when to fold 'em, know when to walk away..." [Used by Gareth Branwyn in *Absolute Beginner's Guide to Building Robots*.]

Recipe for Building Behavior-based Robots
BBR and similar types of bottom-up bots are built in layers, upon the successes of previous layers:

1. Do simple things first.
2. Learn to do them flawlessly.
3. Add new layers of activity over the results of simple tasks.
4. Don't change the simple things.
5. Make new layers work as flawlessly as the previous ones.
6. Repeat ad infinitum.

Common Bonds (and the glues that make them happen)

	Ceramic	Fabric	Glass	Leather	Metal	Paper	Plastic	Rubber	Styrofoam	Vinyl *	Wood
Wood	Epoxy or Fast Epoxy	Spray Adhesive (flexible) or PVA (White Glue)	Silicone-based glue (strong) or PVA (less toxic) or Hot glue (less toxic, fast)	Contact Adhesive or Solvent-free Contact Adhesive	Metal Epoxy (strong) or Contact Cement (large area)	Paste Glue (stays flat, nontoxic) or PVA (strong, less toxic) or Spray (stays flat)	Silicone (strong) or Fast Epoxy (fast) or Contact (large areas)	Contact Adhesive	Polyurethane or Latex Adhesive (strong) or Solvent-free Contact (no clamping) or Hot Glue (fast, less toxic) or PVA (less toxic)	Contact Adhesive or Solvent-free Contact Adhesive	PVA
Vinyl *	Silicone (strong) or Spray (nonreactive)	Spray (stays flat) or Contact	Silicone (strong) or Contact (faster)	Contact (strong) or Spray (large areas) or Solvent-free Contact (less toxic)	Silicone (strong) or Spray (large area)	Spray (stays flat) or Contact	Silicone	Contact or Spray (large area)	Hot Glue (strong, less toxic) or Spray (large areas)	Spray (large area) or Contact	
Styrofoam	Epoxy (strong) or Hot Glue (fast) or PVA (less toxic)	Hot Glue (strong) or PVA or Sprays (invisible)	Epoxy (strong) or Spray (large areas) or PVA (less toxic)	Hot Glue (strong, fast) or Solvent-free Contact (large areas)	Metal Epoxy (strong) or Hot Glue (fast, less toxic)	Spray (stays flat) or PVA or Hot Glue (less toxic)	Epoxy (strong) or Hot Glue (fast, less toxic)	Latex Adhesive	PVA (strong, less toxic) or Solvent-free Contact (carvable, less toxic) or Spray (fast)		
Rubber	Silicone (strong) or Hot Glue (nonreactive)	Contact (strong) or Spray (invisible)	Silicone (strong) or Contact (faster)	Contact or Spray	Silicone (strong) or Spray (large area)	Contact or Spray	Contact (strong) or Spray (large area) or Solvent-free Contact (less toxic)	Spray or Contact			
Plastic	Silicone	Spray	UV Adhesive (if available) or Silicone (strong) or PVA or Hot Glues (less toxic)	Contact	Metal Apoxy	Spray	PVC Adhesive (PVC) or Silicone (flexible) or Fast Epoxy				
Paper	Spray	PVA (less toxic) or Paste (stays flat, nontoxic) or Spray (stays flat)	Spray (large smooth areas) or PVA (less toxic) or Hot Glue (small glass pieces)	Spray (strong) or Paste (nontoxic)	Metal Epoxy (strong) or Spray (large area)	Glue Stick or Paste (stays flat, nontoxic) or PVA (strong, less toxic) or Spray (stays flat)					
Metal	Metal Epoxy (strong) or Cyanoacrylate (invisible)	Spray	UV (if available) or Silicone	Silicone	Metal Epoxy						
Leather	Silicone	Contact (strong) or Spray or PVA (less toxic)	Silicone (strong) or PVA (faster, less toxic)	Contact							
Glass	UV Adhesive (if available) or Silicone	Spray or PVA (large areas) or Hot Glue (small glass pieces, less toxic)	UV Adhesive (if available) or Silicone (strong) or Cyanoacrylate (fast)								

* (treat woven backing as Fabric)
For more info on these bonds and which specific glues to use, go to: thistothat.com

Battery Types and Sizes

Common Name	Other Names	IEC No.*	ANSI/ NEDA No**	Chemical System	Capacity (mAh)	Voltage (V)	Dimensions (mm)		Package Type
							Length/ Height	Diameter/ Width	
A23	23A 3LR50 MN21 Button Stack	3LR50	1181A	Alkaline	65	12	29	10	Cylinder
AA	Penlight Mignon MN1500 MX1500	LR6 R6 FR6 HR6 KR157/51 ZR6	15A 15D 15LF 1.2H2 10015	Alkaline, Carbon-Zinc, Lithium-FeS2, NiMH, NiCd	2700 (Alkaline) 1100 (Carbon-Zinc) 3000 (Lithium-FeS2) 1700–2900 (NiMH) 600–1000 (NiCd)	1.5 1.2 (NiMH and NiCd)	50.5	13.5–14.5	Cylinder
AAA	Micro Microlight MN2400 MX2400	LR03 R03 FR03	24A 24D 24LF	Alkaline, Carbon-Zinc, Li-FeS2	1200 (Alkaline) 540 (Carbon-Zinc) 800–1000 (Ni-MH)	1.5 1.2 (NiMH and NiCd)	44.5	10.5	Cylinder
AAAA	MX2500	LR8D425	25A	Alkaline	625	1.5	42.5	8.3	Cylinder
C	MN1400 MX1400	LR14 R14	14A 14D	Alkaline, Carbon-Zinc, NiMH	8000 (Alkaline) 3800 (Carbon-Zinc) 4500–6000 (NiMH)	1.5 1.2 (NiMH)	50	26.2	Cylinder
CRV3	RCR-V3	N/A	N/A	Lithium, Li-Ion	2000 1300	3	50.8 x 28.57	14.2	Battery Pack
D	Flashlight Battery MN1300 MX1300	LR20 R20	13A 13D	Alkaline, Carbon-Zinc, NiMH	19500 (Alkaline) 8000 (Carbon-Zinc) 9000–11500 (NiMH)	1.5 1.2 (NiMH)	61.5	34.2	Cylinder
E123 (123)	CR123, Lithium Photo	CR17354	5018LC	Lithium, Li-Ion	1500 (Lithium) 700 (Li-Ion)	3	34.5	17	Cylinder
J	7K67	4LR61	1412A	Alkaline	625	6	48.5 x 35.6	9.18	Flat Cartridge
Lantern - Spring Top	6 Volt MN908	4R25Y 4R25	908A 908D	Alkaline, Carbon-Zinc	26000 10500	6	115 x 68.2	68.2	Box
Lantern - Screw Top	6 Volt	4R25Y 4R25	915A 908	Alkaline, Carbon-Zinc	26000 (Alkaline) 10500 (Carbon-Zinc)	6	115 x 68.2	68.2	Box
Lantern - Double Lantern	918 R25-2 Big Lantern MN918	4LR25-24 4R25-2 8R25	918A 918D	Alkaline, Carbon-Zinc	52000 (Alkaline) 22000 (Carbon-Zinc)	6	127 x 136.5	73	Box
N	E90 MN9100	LR1	910A	Alkaline	1000	1.5	30.2	12	Cylinder
9 volt	PP3 MN1604	6LR61 6F22 6KR61	1604A 1604D 1604LC 7.2H5 11604	Alkaline, Carbon-Zinc, Lithium, NiMH, NiCd	565 (Alkaline) 400 (Carbon-Zinc) 1200 (Lithium) 175 (NiMH) 120 (NiCd) 500 (Lithium Polymer)	9 7.2-8.4 (NiMH and NiCd)	26.5 x 48.5	17.5	Box

Battery Types and Sizes

Common Name	Other Names	IEC No.*	ANSI/ NEDA No**	Chemical System	Capacity (mAh)	Voltage (V)	Dimensions (mm)		Package Type
Button and Coin Cell Batteries									
CR927			N/A	Lithium	30	3	2.7	9.5	Coin
CR1220		CR1220	N/A	Lithium	40	3	2.5	12.5	Coin
CR1225		CR1225	N/A	Lithium	50	3	2.5	12.5	Coin
CR1616		CR1616	N/A	Lithium	50	3	1.6	16	Coin
CR1620		CR1620	N/A	Lithium	78	3	2	16	Coin
CR2016	DL2016	CR2016	5000LC	Lithium	90	3	1.6	20	Coin
CR2025	DL2025	CR2025	5003LC	Lithium	160	3	2.5	20	Coin
CR2032	DL2032	CR2032	5004LC	Lithium	225	3	3.2	20	Coin
CR2450	DL2450	CR2450	5029LC	Lithium	610	3	5	24.5	Coin
AC13	DA13H8 Hearing Aid	PR48	7000ZD	Zinc Air	280	1.4	5.4	7.9	Button
SR41	AG3 LR41 D384/392		1135SO 1134SO	Alkaline, Silver Oxide	32 (Alkaline) 42 (Silver Oxide)	1.5 (Alkaline) 1.55 (Silver Oxide)	7.9	3.6	Button
SR43	AG12 LR43 D301/386	LR43 SR43	1133SO 1132SO	Alkaline, Silver Oxide	80 (Alkaline) 120 (Silver Oxide)	1.5 (Alkaline) 1.55 (Silver Oxide)	4.2	11.6	Button
SR44	AG13 LR44 D303/357	LR44 SR44	1166A 1107SO 1131SO	Alkaline, Silver Oxide	150 (Alkaline) 200 (Silver Oxide)	1.50 (Alkaline) 1.55 (Silver Oxide)	5.4	11.6	Button
SR48	AG5 D309/393	SR48	1136SO 1137SO	Silver Oxide	70	1.55	5.4	7.9	Button
SR54	AG10 LR54 D389/390	LR54 SR54	1138SO	Alkaline, Silver Oxide	100 (Alkaline) 70 (Silver Oxide)	1.50 (Alkaline) 1.55 (Silver Oxide)	3.1	11.6	Button
SR55	AG8 D381/391	SR55	1160SO	Silver Oxide	40	1.55	2.1	11.6	Button
SR57	SR927W AG7 D395/399	LR57 SR57	1162SO	Alkaline, Silver Oxide	55	1.55	2.8	9.4	Button
SR58	AG11 D361/362	SR58	1158SO	Silver Oxide	24	1.55	2.1	7.9	Button
SR59	AG2 D396/397	SR59	1163SO	Silver Oxide	30	1.55	2.6	7.9	Button
SR60	AG1 D364	SR60	1175SO	Silver Oxide	20	1.55	2.15	6.8	Button
SR66	AG4 D377	SR66	1176SO	Silver Oxide	26	1.55	2.6	6.8	Button
SR69	AG6 R371	SR69	N/A	Silver Oxide		1.55	2.1	9.5	Button

(*) International Electrotechnical Commission,
(**) American National Standards Institute/National Electronic Distributors Association

Battery Notes:

- The terms "button'" and "coin" batteries are often used synonymously, but coin usually refers exclusively to lithium cells batteries and button to silver oxide (and sometimes alkaline). Both types are also referred to as miniature batteries.

- Alkaline batteries are formulated from manganese dioxide and zinc.

- Silver oxide batteries are also known as silver zinc.

- In button cells, SR denotes silver oxide chemistry, LR denotes manganese dioxide (alkaline).

- When making LED Throwies (see *The Best of MAKE*, pg. 66) use CR2032 coin batteries.

At the margins of precision, the universe wavers.
— *Hu Hsai*

You can go anywhere and do anything if you look important and carry a clipboard.
— *Kata Sutra*

Screw Heads (you should know)

Torx/Star		Common security format for electronics; no longer proprietary (expired patent) but still more expensive; sizes range from T1 to T20+; enhanced-security TR (tamper resistant) version has pin/recess in center
Spanner/Notched		Common security format for hardware in public places; common sizes #4 – 14
Robertson/Square Drive		Format used for hardware in public places, also popular in woodworking; low-slippage; has cult following; tamper-resistant versions have pin in center; common sizes #00 – 3 with color-coded drivers
Tri-Wing		Unusual, expensive security format used by Nintendo; sizes #0 – 4
Torq-Set		Proprietary/expensive high-torque format used in aerospace
PoziDriv/SupaDriv		Proprietary/expensive low-slippage cross-drive format; can be driven with Phillips drivers; popular in Europe
Clutch, Eight-Wing, MorTorq, Quadrex, PoziSquar, Penta-Drive, Spline, Torx Plus, Torx Plus TR, Tri-Groove, Triangle, Triple Square, etc.		For these formats and any others, you can make your own driver bits by hand-machining a hex wrench to shape using a Dremel. If you can't bring the screw into your shop, take a mold of its head with Silly Putty, cast the impression with spackling paste, and use that as your model. See MAKE Volume 03, p. 147.

Dumpster Dives

So, where do dumpster divers go to score? For general finds, big box retailers, stores having closing sales, and apartment complexes. Also high-end neighborhoods the night before large item pickup day, and lightly damaged areas in newly declared disaster zones. For electronics, engineering schools and electronics stores are obvious destinations. Divers will frequently query IT folks about any offices doing widespread upgrades. Construction and demolition sites are where building supplies and materials are found. And manufacturing districts. For instance, glass cutters routinely discard small pieces of glass and mirror. Divers like to scrounge at night and usually carry a stick, a flashlight, or wear a headlamp. For more dirt on diving, visit dumpsterworld.com

Alpha References

C++: Stroustrup; C: Kernighan and Ritchie; Java: Flanagan; Lisp: Steele; Perl: Wall, Christiansen, and Orwant; Python: Lundh; Ruby: Thomas, Fowler, and Hunt; TCL: Raines; Unix: Kernighan and Pike; VB: Lomax. And don't forget: O'Reilly (oreilly.com)

The Ideal Machine is when the function is performed and there is no machine.
— invention methodologist *Genrich Altshuller*

Total Solar Eclipses 2008–2024

1Aug2008: China, Mongolia, Khazakstan, Russia, Greenland, Canada; 22Jul2009: Japan (Southwest Islands), China, Myanmar, Nepal, India; 11Jul2010: Argentina, Chile; 13Nov2012: Australia; 3Nov2013: Somalia, Ethiopia, Kenya, Uganda, Congo, Gabon; 20Mar2015: Norway (Svalbard); 9Mar2016: Indonesia; 21Aug2017: USA; 2Jul2019: Argentina, Chile; 14Dec2020: Argentina, Chile; 4Dec2021: Antarctica; 20Apr2023: Indonesia, Australia; 8Apr2024: Canada, USA, Mexico
Courtesy of Fred Espenak and Sumit Dutta, NASA/GSFC

Mercury Retrogrades 2008–2020

28Jan2008–18Feb, 26May–19Jun, 23Sep–15Oct, 11Jan2009–31Jan, 6May–30May, 6Sep–29Sep, 26Dec–15Jan2010, 17Apr–11May, 20Aug–12Sep, 10Dec–29Dec, 30Mar2011-23Apr, 2Aug–26Aug, 23Nov–13Dec, 11Mar2012–4Apr, 14Jul–7Aug, 6Nov–26Nov, 23Feb2013–17Mar, 26Jun–20Jul, 21Oct–10Nov, 6Feb2014–28Feb, 7Jun–1Jul, 4Oct–25Oct, 21Jan2015–11Feb, 18May–11Jun, 17Sep–9Oct, 5Jan2016–25Jan, 28Apr–22May, 30Aug–21Sep, 19Dec–8Jan2017, 9Apr–3May, 12Aug–5Sep, 2Dec–22Dec, 22Mar2018–15Apr, 25Jul–18Aug, 16Nov–6Dec, 5Mar2019–28Mar, 7Jul–31Jul, 31Oct–20Nov, 16Feb2020–9Mar, 17Jun–12Jul, 13Oct–3Nov

OHM'S LAW

$$I = \frac{V}{R} \qquad V = I \times R \qquad R = \frac{V}{I}$$

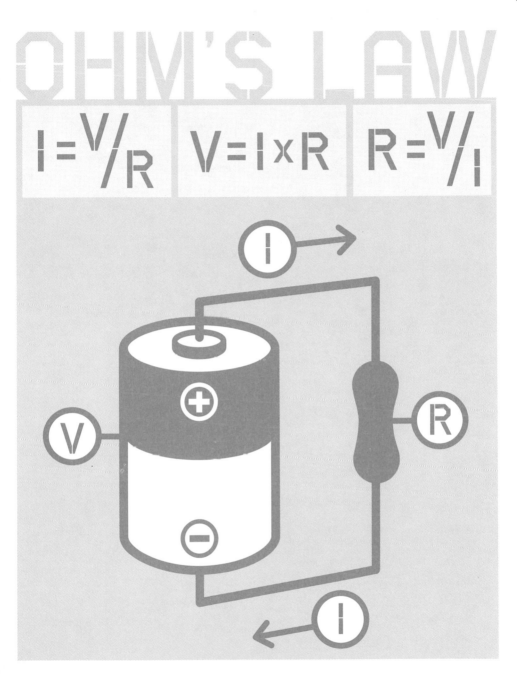

If a known current (I) flows through a resistor (R), voltage (V) can be calculated:
V = I x R

If you know the voltage (V) across a resistor (R), current (I) can be calculated:
I = V / R

If current (I) flows through a resistor, and there is a voltage across the resistor (R), resistance can be calculated: **R = V / I**

(So, if R stands for resistance, and V for volts, what does the I for current stand for? Bet you didn't guess *intensity*.)

Basic Electronics Components and Function

Component	Symbol	Function	Measurement/ Classification	Notes
Fixed Resistor		Restricts the flow of current in a circuit to a set value	Ohms (Ω)	Also measured in kΩ (1000) and MΩ (1 million)
Variable Resistor		Restricts the flow of current in a circuit to a range of values	Ohms (Ω)	A popular type of variable resistor is called a potentiometer (or "pot") which uses a sliding contact or control dial
Light-dependent Resistor		A type of resistor that converts light to electrical resistance	Ohms (Ω) and Lux	The resistance of an LDR decreases as the level of light increases
Thermistor		A type of resistor that changes its resistance value in response to temperature changes	PTC (Positive Temperature Coefficient) and NTC (Negative Temperature Coefficient)	In a PTC thermistor, resistance increases with increasing temperature, in an NTC, resistance decreases with increasing temperature
Capacitor		Temporarily stores current in a circuit and can also be used in filtering a signal	Farads (F)	Also measured μF (millionths). Can be either polarity-sensitive or non-polarity sensitive (fixed).
Diode	Anode ▷⊢ Cathode	Restricts current flow to one direction only	A 1N-series numbering scheme (e.g., 1N34A/ 1N270 (Germanium signal), IN914/1N4148 (Silicon signal), and 1N4001–1N4007 (Silicon 1A power rectifier)	There are a variety of diodes, including the most common (rectifying), Zener (which can conduct backwards at a set voltage), and Schottky diodes (useful in voltage clamping)
Light Emitting Diode (LED)	Anode ▷⊢ Cathode	A type of diode that produces light when powered. Used in everything from status indicator lights to flashlights and room lighting.	Measured/classed by Amps (mA), Volts, Candela (mcd), package size (e.g., 2mm, 3mm, 5mm), color, shape	The cathode (–) side of the LED package is usually notched or flat. LEDs are available in light ranges from infrared to ultraviolet.
Transistor	NPN PNP	A type of semiconductor used to act as a switch in a circuit or to amplify a signal	Measured by structure (BJT, JFET, MOSFET), polarity, and power rating (low, med, high)	One of the most common types of transistor is the bi-polar junction transistor (BJT) which comes in two polarities (PNP and NPN) and has three pins (Emitter, Base, Collector)
Switch	L1 L2 COM	A mechanical or electronic device used to open, close or connect portions of an electrical circuit	Poles (numbers of circuits to control) and Throws (numbers of circuit paths that are controllable)	Common switch types are designated as SPST (single-pole, single throw), i.e., your average wall switch; SPDT, i.e., a switch that can switch between two circuits; and DPDT, i.e., a switch that can switch two circuits simultaneously
Relay		A device that creates an electromagnetic field used to make (or break) an electrical contact (or contacts)	Pole and Throws, and Volts	A diagram showing the switch pins and the orientation of the electromagnet is frequently printed right on the relay package
Integrated Circuit	Vcc NC A Q B IC NC C Vdd NC	A miniaturized electronic circuit that has been embedded in a substrate of semiconductor material	Type (analog, digital, hybrid), number of pins, function, chip family	IC pins are identified by starting at the one with the dimple (top right with the pins facing down) being pin 1, counting down the left side, moving across to the bottom most right pin and counting up
Battery	+ −	Used to store and deliver DC power to a circuit	Letter size (AAA, AA, C), ampere-hour (Ah), Volts (V)	An array of electrochemical cells is called a battery. The polarity of most batteries is clearly marked and on most, the (+) side is a nipple/nub.
Motor	—(M)—	An electric motor is a device used to convert electromagnetic energy into mechanical motion	Volts, Amps, RPM, and torque	Common types of DC motors include brushed and brushless, the stepper, and the servo-motor (which is a control mechanism coupled with a DC gearmotor)
Transformer		A device for transferring electrical energy from one circuit to another by means of a magnetic field shared by the two circuits	Many: Volt/amps, frequency, cool type, winding ratio (step-up, step down), application and purpose (power supply, current stabilizer, rectifier, etc.)	
Lamp	—⊗—	A device that transforms electrical energy into light	Watts, Candela, Lumen, Lux	On circuit diagrams, an "X" through the lamp symbol usually means that it is to be used as an indicator light
Wire		A conductive material used to carry an electrical current. Usually covered in an insulating material called a jacket.	Measured in gauges, such as the American Wire Gauge (AWG)	In circuit diagrams, wires that connect are usually indicated by a dot on the junction of the wires

Basic DMM Circuit Tests

BASIC DMM CIRCUIT TESTS

A Digital Multimeter (DMM) is a tool that every maker should own and know how to use. This chart outlines the basic tests that can be performed on AC and DC circuits (and individual components). These images show the general procedures, NOT the specifics for your multimeter. You must check the manual that came with your tool to find the right settings and test procedures.

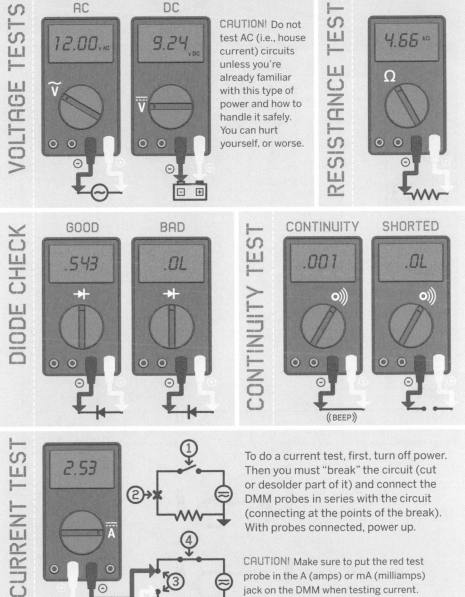

VOLTAGE TESTS

AC — 12.00 v AC

DC — 9.24 v DC

CAUTION! Do not test AC (i.e., house current) circuits unless you're already familiar with this type of power and how to handle it safely. You can hurt yourself, or worse.

RESISTANCE TEST

4.66 kΩ — Ω

DIODE CHECK

GOOD — .543

BAD — .OL

CONTINUITY TEST

CONTINUITY — .001 — ((BEEP))

SHORTED — .OL

CURRENT TEST

2.53 — A

To do a current test, first, turn off power. Then you must "break" the circuit (cut or desolder part of it) and connect the DMM probes in series with the circuit (connecting at the points of the break). With probes connected, power up.

CAUTION! Make sure to put the red test probe in the A (amps) or mA (milliamps) jack on the DMM when testing current.

Resistor Codes

RESISTOR COLOR CODES

STANDARD FOUR
BAND RESISTOR

COLOR	1ST	2ND	3RD	MULTIPLIER	TOLERANCE	RELIABILITY
	A	B	C	D	E	F
BLACK		0	0	1		
BROWN	1	1	1	10	± 1%	1%
RED	2	2	2	10^2	± 2%	0.1%
ORANGE	3	3	3	10^3	± 3%	0.01%
YELLOW	4	4	4	10^4	± 4%	0.001%
GREEN	5	5	5	10^5	± 0.5%	
BLUE	6	6	6	10^6	± 0.25%	
VIOLET	7	7	7	10^7	± 0.1%	
GREY	8	8	8	10^8		
WHITE	9	9	9	10^9		
GOLD					± 5%	
SILVER					± 10%	
NONE					± 20%	

A B C D E F

FIVE BAND
(PRECISION) RESISTOR

FIVE BAND
(RELIABILITY)
RESISTOR*

*Found in military electronics

Capacitors

Figuring out the values of capacitors from the strange numbers, symbols, and colors on them can be a bit of an arcane science. We didn't bother covering color-coded caps here, as those are not that common any longer. When in doubt: google.com.

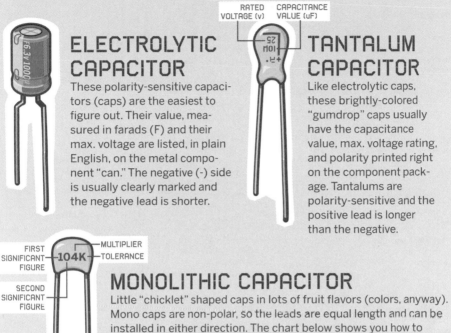

RATED VOLTAGE (v)   CAPACITANCE VALUE (uF)

ELECTROLYTIC CAPACITOR

These polarity-sensitive capacitors (caps) are the easiest to figure out. Their value, measured in farads (F) and their max. voltage are listed, in plain English, on the metal component "can." The negative (-) side is usually clearly marked and the negative lead is shorter.

TANTALUM CAPACITOR

Like electrolytic caps, these brightly-colored "gumdrop" caps usually have the capacitance value, max. voltage rating, and polarity printed right on the component package. Tantalums are polarity-sensitive and the positive lead is longer than the negative.

FIRST SIGNIFICANT FIGURE — **104K** — MULTIPLIER / TOLERANCE

SECOND SIGNIFICANT FIGURE

MONOLITHIC CAPACITOR

Little "chicklet" shaped caps in lots of fruit flavors (colors, anyway). Mono caps are non-polar, so the leads are equal length and can be installed in either direction. The chart below shows you how to determine their value using the three-number plus letter code found on most monolithic caps.

MONOLITHIC CAPACITOR CHART

VALUE (Fig 1 & 2)	MULTIPLIER	LETTER	TOLERANCE
0	1	B	± 0.1pF
1	10	C	± 0.25pF
2	10^2	D	± 0.5pF
3	10^3	F	± 1%
4	10^4	G	± 2%
5	10^5	H	± 3%
6	N/A	J	± 5%
7	N/A	K	± 10%
8	0.01	M	± 20%
9	0.1	Z	± 80%/-20%

CAPACITOR CODEBREAKER

COMMON CAPACITOR CODE	PICOFARAD (pF)	NANOFARAD (nF)	MICROFARAD (mF,uF or mfd)
102	1000	1 or 1n	0.001
152	1500	1.5 or 1n5	0.0015
222	2200	2.2 or 2n2	0.0022
332	3300	3.3 or 3n3	0.0033
472	4700	4.7 or 4n7	0.0047
682	6800	6.8 or 6n8	0.0068
103	10000	10 or 10n	0.01
153	15000	15 or 15n	0.015
223	22000	22 or 22n	0.022
333	33000	33 or 33n	0.033
473	47000	47 or 47n	0.047
683	68000	68 or 68n	0.068
104	100000	100 or 100n	0.1
154	150000	150 or 150n	0.15
224	220000	220 or 220n	0.22
334	330000	330 or 330n	0.33
474	470000	470 or 470n	0.47

LED Color Chart

Color Name	Wavelength (nm)	Fwd Voltage (Vf @ 20ma)	Intensity 5mm LEDs	Viewing Angle	LED Dye Material
High Efficiency Red	640	2	220mcd @20mA	12°	GaAsP/GaP – Gallium Arsenic Phosphide/Gallium Phosphide
Super Red	634	2.2	8000mcd @20mA	12°	InGaAlP – Indium Gallium Aluminum Phosphide
Red-Orange	623	2.2	4500mcd @20mA	12°	InGaAlP – Indium Gallium Aluminum Phosphide
Orange	609	2.1	220mcd @20mA	12°	GaAsP/GaP – Gallium Arsenic Phosphide/Gallium Phosphide
Super Yellow	598	2.1	500mcd @20mA	12°	InGaAlP – Indium Gallium Aluminum Phosphide
Yellow	582	2.1	170mcd @20mA	12°	GaAsP/GaP – Gallium Arsenic Phosphide/Gallium Phosphide
Warm White	3000K	3.6	5500mcd @20mA	12°	SiC/GaN – Silicon Carbide/Gallium Nitride
Pale White	6000K	3.6	5500mcd @20mA	40°	SiC/GaN – Silicon Carbide/Gallium Nitride
Cool White	8000+K	3.6	5800mcd @20mA	12°	SiC/GaN – Silicon Carbide/Gallium Nitride
Super Lime Yellow	575	2.4	1800mcd @20mA	12°	InGaAlP – Indium Gallium Aluminum Phosphide
Super Lime Green	575	2	1800mcd @20mA	12°	InGaAlP – Indium Gallium Aluminum Phosphide
High Efficiency Green	563	2.1	210mcd @20mA	12°	GaP/GaP – Gallium Phosphide/Gallium Phosphide
Super Pure Green	560	2.1	350mcd @20mA	40°	InGaAlP – Indium Gallium Aluminum Phosphide
Pure Green	555	2.1	140mcd @20mA	12°	GaP/GaP – Gallium Phosphide/Gallium Phosphide
Aqua Green	525	3.5	10,000mcd @20mA	12°	SiC/GaN – Silicon Carbide/Gallium Nitride
Blue Green	501	3.5	4300mcd @20mA	40°	SiC/GaN – Silicon Carbide/Gallium Nitride
Super Blue	455	3.6	3000mcd @20mA	12°	SiC/GaN – Silicon Carbide/Gallium Nitride
Ultra Blue	425	3.8	250mcd @20mA	12°	SiC/GaN – Silicon Carbide/Gallium Nitride

Source: LEDTronics.com. Used with permission.

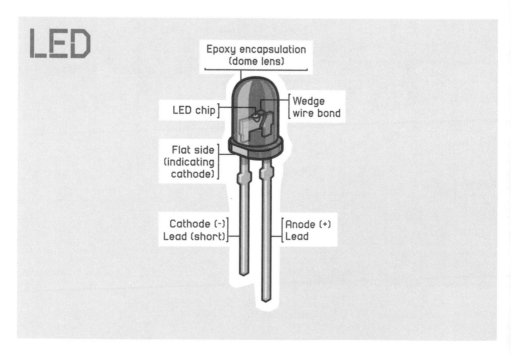

American Wire Gauge

AWG Size	Conductor Diameter		Cross Section (Copper Area)		Resistance (Copper)		Max Amps**	
	(inch)	(mm)	(kcmil)	(mm²)	(Ω/km)	(Ω/1000 ft)	Chassis Wiring	Power Transmission
0000 (4/0)	0.46	11.7	212	107	0.16*	0.049*	380	302
000 (3/0)	0.41	10.4	168	85	0.2*	0.062*	328	239
00 (2/0)	0.365	9.27	133	67.4	0.25*	0.077*	283	190
0 (1/0)	0.325	8.25	106	53.5	~0.3281	~0.1	245	150
1	0.289	7.35	83.7	42.4	0.4*	0.12*	211	119
2	0.258	6.54	66.4	33.6	0.5*	0.15*	181	94
3	0.229	5.83	52.6	26.7			158	75
4	0.204	5.19	41.7	21.2	0.8*	0.24*	135	60
5	0.182	4.62	33.1	16.8			118	47
6	0.162	4.12	26.3	13.3	1.5*	0.47*	101	37
7	0.144	3.66	20.8	10.5			89	30
8	0.128	3.26	16.5	8.37	2.2*	0.67*	73	24
9	0.114	2.91	13.1	6.63			64	19
10	0.102	2.59	10.4	5.26	3.2772	0.9989	55	15
11	0.0907	2.3	8.23	4.17	4.1339	1.26	47	12
12	0.0808	2.05	6.53	3.31	5.21	1.588	41	9.3
13	0.072	1.83	5.18	2.62	6.572	2.003	35	7.4
14	0.0641	1.63	4.11	2.08	8.284	2.525	32	5.9
15	0.0571	1.45	3.26	1.65	10.45	3.184	28	4.7
16	0.0508	1.29	2.58	1.31	13.18	4.016	22	3.7
17	0.0453	1.15	2.05	1.04	16.614	5.064	19	2.9
18	0.0403	1.02	1.62	0.823	20.948	6.385	16	2.3
19	0.0359	0.912	1.29	0.653	26.414	8.051	14	1.8
20	0.032	0.812	1.02	0.518	33.301	10.15	11	1.5
21	0.0285	0.723	0.81	0.41	41.995	12.8	9	1.2
22	0.0253	0.644	0.642	0.326	52.953	16.14	7	0.92
23	0.0226	0.573	0.509	0.258	66.798	20.36	4.7	0.729
24	0.0201	0.511	0.404	0.205	84.219	25.67	3.5	0.577
25	0.0179	0.455	0.32	0.162	106.201	32.37	2.7	0.457
26	0.0159	0.405	0.254	0.129	133.891	40.81	2.2	0.361
27	0.0142	0.361	0.202	0.102	168.865	51.47	1.7	0.288
28	0.0126	0.321	0.16	0.081	212.927	64.9	1.4	0.226
29	0.0113	0.286	0.127	0.0642	268.471	81.83	1.2	0.182
30	0.01	0.255	0.101	0.0509	338.583	103.2	0.86	0.142
31	0.00893	0.227	0.0797	0.0404	426.837	130.1	0.7	0.113
32	0.00795	0.202	0.0632	0.032	538.386	164.1	0.53	0.091
33	0.00708	0.18	0.0501	0.0254	678.806	206.9	0.43	0.072
34	0.0063	0.16	0.0398	0.0201	833	260.9	0.33	0.056
35	0.00561	0.143	0.0315	0.016	1085.958	331	0.27	0.044
36	0.005	0.127	0.025	0.0127	1360.892	414.8	0.21	0.035
37	0.00445	0.113	0.0198	0.01	1680.118	512.1	0.17	0.0289
38	0.00397	0.101	0.0157	0.00797	2127.953	648.6	0.13	0.0228
39	0.00353	0.0897	0.0125	0.00632	2781.496	847.8	0.11	0.0175
40	0.00314	0.0799	0.00989	0.00501	3543.307	1080	0.09	0.0137

(*) Insulation included
(**) Rule of thumb rating only. Careful engineering and testing required for application.

AWG: In American Wire Gauge (AWG), diameters can be calculated by applying the formula $D(AWG)=.005\cdot92^{((36-AWG)/39)}$ inch. For the 00, 000, 0000, etc., gauges you use -1, -2, -3, which makes more sense mathematically than "double-ought." This means that in AWG, every six-gauge decrease gives a doubling of the wire diameter, and every three-gauge decrease doubles the wire cross-sectional area. Similar to dB in signal and power levels. From Powerstream.com (powerstream.com/Wire_Size.htm)

Metric Gauge: In the Metric Gauge scale, the gauge is 10 times the diameter in millimeters, so a 50 gauge metric wire would be 5 mm in diameter. Note that in AWG the diameter goes up as the gauge goes down, but for metric gauges it is the opposite. Probably because of this confusion, most of the time metric-sized wire is specified in millimeters rather than metric gauges. From Powerstream.com (powerstream.com/Wire_Size.htm)

Common Pinouts

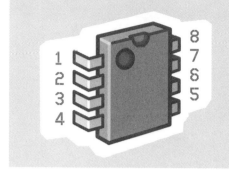

Here's a selection of pinouts for some of our "favorite" chips and connectors. Doing a web search on the name of the chip will return a wealth of information, such as the manufacturer's datasheet.

General Wiring Colors
These are the commonly used color-codes, though not all manufacturers follow them. Pinout diagrams are usually more reliable.

Black = GND
Orange = +3.3v
Red = +5v
Yellow = +12v
White = -5v
Blue = -12v

USB
Serial Bus
Pin 1: Vcc (Red)
Pin 2: D- (Data) (White)
Pin 3: D+ (Green)
Pin 4: GND (Black)

1381
Voltage Detector
(TO-92 package)
Pin 1: OUT
Pin 2: V+
Pin 3: GND

78L0*
Voltage Regulator
Pin 1: OUT
Pin 2: COM
Pin 3: COM
Pin 4: NC
Pin 5: NC
Pin 6: COM
Pin 7: COM
Pin 8: IN

LM2937
Low Drop-out Voltage Regulator
Pin 1: IN
Pin 2: GND
Pin 3: OUT

LM325
Dual Regulator
Pin 1: +Boost
Pin 2: NC
Pin 3: +Vin
Pin 4: -Vin
Pin 5: -Current Limit
Pin 6: -Sense
Pin 7: -Vout
Pin 8: -Boost
Pin 9: NC
Pin 10: Reference
Pin 11: GND
Pin 12: NC
Pin 13: +Current Limit
Pin 14: +Sense

MAX8211
2-sided voltage detector
Pin 1: NC
Pin 2: Hysteresis
Pin 3: Threshold
Pin 4: OUT
Pin 5: GND
Pin 6: NC
Pin 7: NC
Pin 8: V+

LM324
Quad OpAmp
Pin 1: OUT1
Pin 2: IN1
Pin 3: +IN1
Pin 4: Vcc
Pin 5: +IN2
Pin 6: -IN2
Pin 7: OUT2
Pin 8: OUT3
Pin 9: -IN3
Pin 10: +IN2
Pin 11: GND
Pin 12: +IN4
Pin 13: -IN4
Pin 14: OUT4

LM386
OpAmp
Pin 1: Gain
Pin 2: -IN
Pin 3: +IN
Pin 4: GND
Pin 5: V out
Pin 6: Vs
Pin 7: Bypass
Pin 8: Gain

741
OpAmp
Pin 1: Offset Null
Pin 2: -IN
Pin 3: +IN
Pin 4: -V
Pin 5: Offset Null
Pin 6: OUT
Pin 7: +V
Pin 8: Strobe

TL081
JFET Input OpAmp
Pin 1: Balance
Pin 2: - IN
Pin 3: + IN
Pin 4: V-
Pin 5: Balance
Pin 6: OUT
Pin 7: V+
Pin 8: NC

555
Timer
Pin 1: GND
Pin 2: Trigger
Pin 3: Out
Pin 4: Reset
Pin 5: Control Voltage
Pin 6: Threshold
Pin 7: Dicharge
Pin 8: +5V

556
Dual Timer
Pin 1: Discharge A
Pin 2: Threshold A
Pin 3: Control A
Pin 4: Reset A
Pin 5: OUT A
Pin 6: Trigger A
Pin 7: 0v
Pin 8: Trigger B
Pin 9: OUT B
Pin 10: Reset B
Pin 11: Control B
Pin 12: Threshold B
Pin 13: Discharge B
Pin 14: +4.5 - 15V

L293B/D
Dual H-Bridge Motor Driver
Pin 1: Enable 1
Pin 2: IN 1
Pin 3: OUT 1
Pin 4: GND
Pin 5: GND
Pin 6: OUT 2
Pin 7: IN 2
Pin 8: Motor Power
Pin 9: Enable
Pin 10: IN 3
Pin 11: OUT 3
Pin 12: GND
Pin 13: GND
Pin 14: OUT 4
Pin 15: IN 4
Pin 16: Logic Power

SN754410
Quad Half-H driver
Pin 1: Enable 1, 2
Pin 2: IN1
Pin 3: OUT1
Pin 4: Heat Sink & GND
Pin 5: Heat Sink & GND
Pin 6: OUT2
Pin 7: IN2
Pin 8: Vcc2
Pin 9: Enable 3, 4
Pin 10: IN3
Pin 11: OUT3
Pin 12: Heat Sink & GND
Pin 13: Heat Sink & GND
Pin 14: OUT4
Pin 15: IN4
Pin 16: Vcc1

74AC240
Octal Buffer
Pin 1: Enable 1
Pin 2: IN0
Pin 3: OUT4
Pin 4: IN1
Pin 5: OUT5
Pin 6: IN2
Pin 7: OUT6
Pin 8: IN3
Pin 9: OUT7
Pin 10: GND
Pin 11: IN7
Pin 12: OUT3
Pin 13: IN6
Pin 14: OUT2
Pin 15: IN5
Pin 16: OUT1
Pin 17: IN4
Pin 18: OUT0
Pin 19: Enable 2
Pin 20: Vcc

AVR ISP
In-System Programming Connector
Pin 1: MOSI
Pin 2: V+
Pin 3: LED
Pin 4: GND
Pin 5: Reset
Pin 6: GND
Pin 7: SCK
Pin 8: GND
Pin 9: MISO
Pin 10: GND

ATtiny25
(and ATtiny 45,
ATtiny85)
8-bit AVR
Microcontroller
Pin 1: PCINT5/Reset/ADC0/dW
Pin 2: PCINT3/XTAL1/CLK1/ADC3
Pin 3: PCINT4/XTAL2/CLK0/OC1B/ADC2
Pin 4: GND
Pin 5: MOSI/DI/SDA/AIN0/OC0A/OC1A/AREF/PCINT0
Pin 6: MISO/DO/AIN1/OC0B/OC1A/PCINT1
Pin 7: SCK/DO/USCK/SCL/ADC1/T0/INT0/PCINT2
Pin 8: Vcc

ATMega 8
Microcontroller
(used in Arduino)
Pin 1: Reset
Pin 2: RXD
Pin 3: TXD
Pin 4: INT0
Pin 5: INT1
Pin 6: XCK/T0
Pin 7: Vcc
Pin 8: GND
Pin 9: XTAL1/TOSC1
Pin 10: XTAL2/TOSC2
Pin 11: T1
Pin 12: AIN0
Pin 13: AIN1
Pin 14: ICP1
Pin 15: OC1A
Pin 16: SS/OC1B
Pin 17: MOSI/OC2
Pin 18: MISO
Pin 19: SCK
Pin 20: AVcc
Pin 21: AREF
Pin 22: GND
Pin 23: ADC0
Pin 24: ADC1
Pin 25: ADC2
Pin 26: ADC3
Pin 27: ADC4/SDA
Pin 28: ADC5/SCL

ATtiny2313
Microcontroller
Pin 1: Reset/dW
Pin 2: RXD
Pin 3: TXD
Pin 4: XTAL2
Pin 5: XTAL1
Pin 6: CKOUT/XCK/INT0
Pin 7: INT1
Pin 8: T0
Pin 9: OC0B/T1
Pin 10: GND
Pin 11: ICP
Pin 12: AIN0/PCINT0
Pin 13: AIN1/PCINT1
Pin 14: OC0A/PCINT2
Pin 15: OC1A/PCINT3
Pin 16: OC1B/PCINT4
Pin 17: MOSI/DI/SDA/PCINT5
Pin 18: MISO/DO/PCINT6
Pin 19: UCSK/SCL/PCINT7
Pin 20: Vcc

ATMega48
(and ATMega 88,168,328)
Microcontroller
(used in Arduino)
Pin 1: PCINT14/Reset
Pin 2: PCINT16/RXD
Pin 3: PCINT17/TXD
Pin 4: PCINT18/INT0
Pin 5: PCINT19/OC2B/INT1
Pin 6: PCINT20/XCK/T0
Pin 7: Vcc
Pin 8: GND
Pin 9: PCINT6/XTAL1/TOSC1
Pin 10: PCINT7/XTAL2/TOSC2
Pin 11: PCINT21/OC0B/T1
Pin 12: PCINT22/OC0A/AIN0
Pin 13: PCINT23/AIN1
Pin 14: PCINT0/CLKO/ICP1
Pin 15: OC1A/PCINT1
Pin 16: SS/OC1B/PCINT2
Pin 17: MOSI/OS2A/PCINT3
Pin 18: MISO/PCINT4
Pin 19: SCK/PCINT5
Pin 20: AVcc
Pin 21: AREF
Pin 22: GND
Pin 23: ADC0/PIINT8
Pin 24: ADC1/PCINT9
Pin 25: ADC2/PCINT10
Pin 26: ADC3/PCINT11
Pin 27: ADC4/SDA/PCINT12
Pin 28: ADC5/SCL/PCINT13

Special Thanks

Maker's Notebook Brain Trust

Mark Frauenfelder
Brian Jepson
Phillip Torrone
Windell Oskay
Lenore Edman
Jeffrey McGrew
Jillian Northrup
Jake von Slatt
Richard Nagy ("Datamancer")
Natalie Zee Drieu
Michelle Kempner
Becky Stern
Patti Schiendelman
Sean Carton
R. Mark Adams
Perry Kaye
Shawn Connally
Terrie Miller
Goli Mohammadi
Ty Nowotny
Rob Bullington
Eric Michael Beug
Jason Striegel
Jonah Brucker-Cohen
Collin Cunningham

Acknowledgements

Judy Willard
Katie Gekker
Fannon Printing
Mr. Jalopy
Ulla-Maaria Mutanen
Alberto Gaitán
thistothat.com
LEDtronics.com

The best way to have a good idea is to have lots of ideas. — *Linus Pauling*

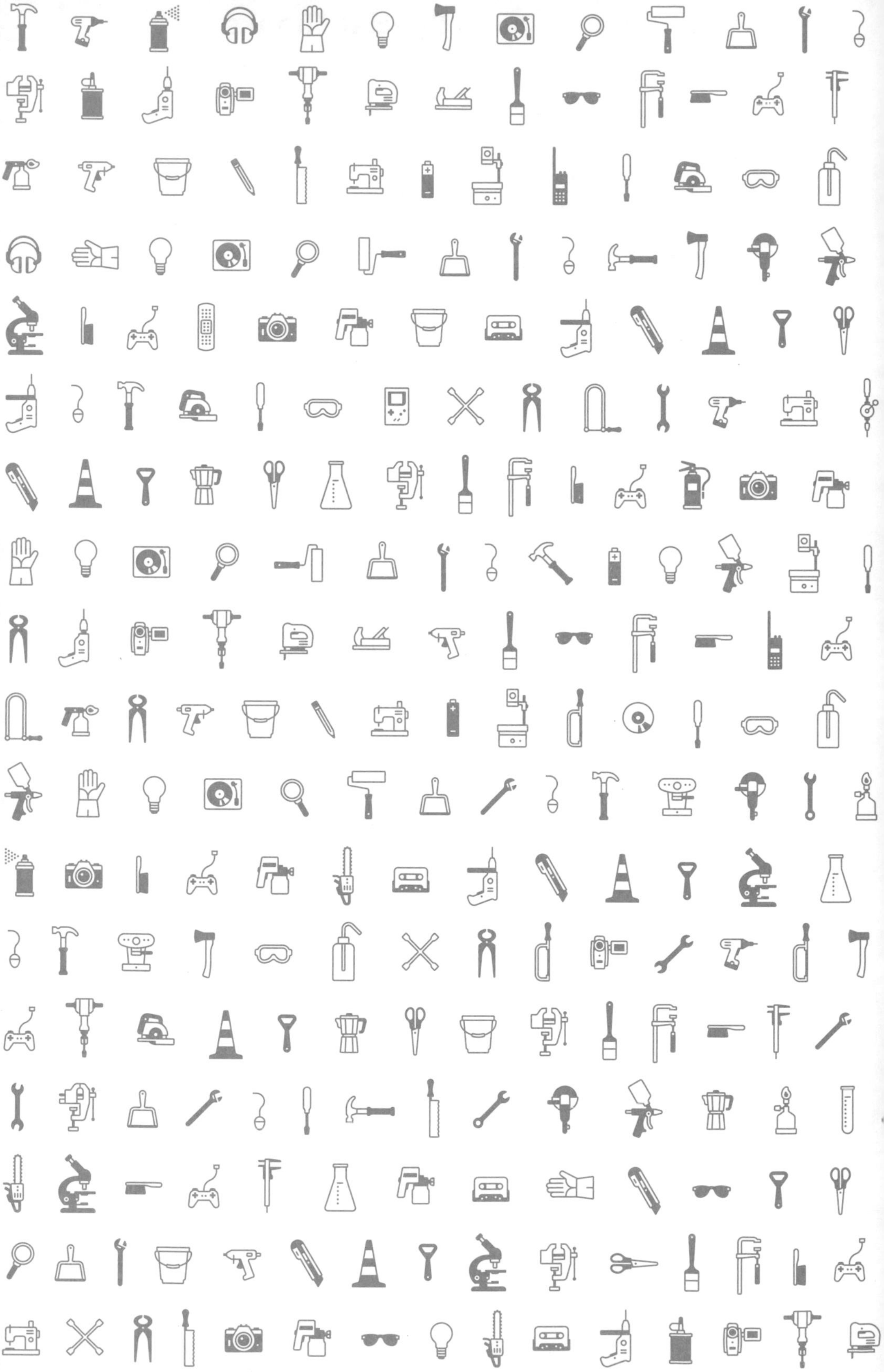

Make: